Introduction

When I first became interested in le [illegible] on the Friday morning walks led by VI National Park Ranger, Laurel [illegible] Francis Bay. That's a great place to look for birds because it has three types of areas where different birds can be found: the woods, the pond, and the beach.

At first it was hard for me to see some of the birds she pointed out. Binoculars helped, though it was a bit difficult to figure out how to use them the right way. And then the birds often flew off before I could focus on them. However, sometimes they cooperated, and I was grateful for that.

Once I learned more about how to find and recognize birds, I discovered that I really liked looking. When I saw movement in a tree, or heard a song, I learned to be very quiet and still, and just pay attention. It was calming, but then also exciting like playing a challenging game.

After I got a camera with a long lens, I could see small birds more clearly, and was able to identify them using the photos I took. I would look at books or online resources to figure out what birds they were. Or ask Laurel or other experts.

By now, I have gotten to know quite a few birds on St. John, and taken lots of photos, which I am happy to share here. These photos do not represent all the birds that can be seen on St. John, but they do include the most common ones. Some birds are here all the time, some fly down during the winter when there isn't enough food for them in North America, and others come over from other places nearby, or just pass through briefly when they are migrating.

Many of the birds in this book are ones you can see around the house or along the road. Others you have to go look for. Once you learn to recognize the local birds, and what sounds they make, you can feel like you have new friends and neighbors. You might hear them calling, and then look up and say: "Those are White-winged Doves on the wire" or "There's a group of Smooth-billed Anis flying through the trees".

When the birds feel like your friends, then you might start thinking about what their lives are like here, and even want to help look out for them. You could set up feeders for the bananaquits and the hummingbirds when the weather is dry and there aren't many flowers around with nectar for them to eat. Or you could put out seeds for the doves and bullfinches. Or plant trees with berries, seeds and flowers that different birds like.

It's been fun for me to have birds become part of my life. The birds help me feel more connected with the natural world around me, including the plants and trees, ponds and woodlands. I often feel happy just because I see a familiar bird come by. I hope you will feel some of that happiness too as you learn more about birds on St. John.

Gail Karlsson
VI Audubon Society Board Member

The Virgin Islands Audubon Society

Our Mission

- To foster an appreciation of the wildlife existing in and around the VI and the Caribbean

- To actively pursue protection of habitat in which the wildlife exists, and prevent and reverse the deterioration of habitat wherever possible

- To study the behavioral patterns of the wildlife

- To educate the public as well as ourselves about the wildlife and their habitats, and

- To raise funds to serve the nonprofit protective and educational purposes of the Society

The VI Audubon Society is a 501)(c)(3) nonprofit organization organized exclusively for charitable, educational and scientific purposes, including making distributions to organizations.

Virgin Islands Audubon Society, Estate Enighed, PMB 402, St. John, Virgin Islands 00830

Facebook: Virgin Islands Audubon Society
Email: viasbirds@gmail.com

Woodland Birds

Anis

Smooth-billed Anis are resident birds with big bills like parrots, but they are actually related to cuckoos. They are sometimes called 'black witches'. They usually fly around in the tops of trees in small flocks of 5 or 6, looking for large insects, including the caterpillars that eat the leaves off the Frangipani trees. The members of the flock even share the same nest. When they call to each other it sounds like their name: "*ah-nee, ah-nee*".

Bananaquits

Bananaquits are local favorites, also called Yellow Breasts, or Sugar Birds because they will eat from sugar feeders. They have long tongues they use to get nectar from flowers, and to drink sugar water. Otherwise they eat fruits, berries and small bugs. They will sometimes make their nests in a cactus for protection.

Bullfinches

Lesser Antillean Bullfinches mostly eat seeds and berries, but will also eagerly come to sugar feeders. The males are black with red patches, and the females are greenish tan. They are resident birds that nest in trees, and sometimes in rafters of houses.

Male

Female

Cuckoos

Mangrove Cuckoos are shy but beautiful local birds that live in the woods around the wetlands. They mostly stay hidden as they hunt for large insects, lizards and berries. They have a long, drawn-out call *"dat, dat, dat, dat, dat, dat, dat"* that drops in tone and sometimes makes people wonder if there are monkeys hiding in the bush. There are also occasionally Yellow-billed Cuckoos that pass through St. John during spring or fall migration, and they look very similar.

Doves

Bridled Quail-Doves are mostly found walking around on the ground in moist forest areas, looking for seeds and berries. Their name refers to the white stripes on their faces, like when a rider puts a bridle in a horse's mouth. You might see them at the bottom of the Reef Bay trail, or across the road from Cinnamon Bay. They are the largest doves living on St. John, about a foot long.

Common Ground Doves can be seen foraging on island roads and trails, including the Francis Bay trail, as well as in open grassy areas. You might also see them perched in trees. They are usually together as a couple or in a small group, and are the smallest of the doves living on St. John, only about 5 to 6 inches long.

Eurasian Collared-Doves have distinctive black marks on the backs of their necks. They are relatively rare here, though there are more in Puerto Rico and some of the British Virgin Islands. They only recently came over to St. John and are mostly in the Frank Bay area near town. They are native to Europe and Asia.

White-winged Doves are also quite recent residents of St. John. They started moving over from Puerto Rico about 25 years ago. When they are sitting down you can easily see the white edges on their wings. They can be very noisy, sitting on treetops or wires calling out to each other with repeated sounds like "*who cooks for you*" and "*I wanted you*".

Zenaida Doves like to walk along the dirt roads looking for seeds on the ground. When a car comes by, they fly up at the last second, and their wings make a whistling sound. At other times they sit in the trees and call to each other: "*who are you, you, you*". When their throats expand as they call, you can see the purple iridescent feathers there. They look like Mourning Doves in North America, but have distinctive little patches of white on their wing tips.

Elaenias

Caribbean Eleanias are native flycatchers, and are hard to see because they hide in among the leaves of trees looking for bugs and small fruit. But even if you can't see them, you might hear one calling out loudly, over and over again.

Falcons

American Kestrels are resident falcons that perch on top of dead trees or on wires along the road. From there they can spot lizards and mice, then drop down to grab their prey. The males have blue on the edges of their wings, and the females' wings are mostly brown. They use holes in trees for their nests, or sometimes niches in buildings.

Merlins are fast-flying falcons that occasionally come to St. John in the fall or winter. They are larger than the American Kestrels, and mostly eat smaller birds, which they can catch in mid-air. During one bird walk, a Merlin was hiding behind a branch of a tall tree near the Francis Bay trail.

Peregrine Falcons are rare on St. John but do show up sometimes in the wetlands or high forests during the winter. They are larger than Merlins, and have thick dark streaks on their faces. They have long, thin wings, and can chase other birds at speeds of over 200 miles per hour, then knock them out of the sky and capture them.

Grassquits

Black-faced Grassquits are small residents that mostly eat seeds and bugs they find on the ground or in tree branches. They often forage together as a couple. The male is the darker one, while the female is a greenish tan.

Male

Female

Hawks

Red-tailed Hawks are sometimes called 'chicken hawks'. They are large, powerful raptors that live full time on St. John. They can often be seen soaring high overhead, riding currents of rising air heated by the sun. They will drop down to catch rodents, lizards, crabs, large insects and smaller birds. They nest high up on the hilltops.

Hummingbirds

Male

Antillean Crested Hummingbirds are tiny local birds with thin bills designed for drinking nectar from flowers. Their wings beat very fast so they can hover over the flowers while they are drinking. The females make tiny nests attached to small tree branches. They have iridescent feathers that shine in the sunlight, and the males have crest feathers on their heads that light up like a green flash when the sun's rays hit them right.

Female

Green-throated Carib hummingbirds are larger than the Antillean Crested ones, with longer, curved bills. Both types of hummingbirds are nectar drinkers, and will come to sugar water feeders, aggressively chasing away other birds. While they drink nectar from one flower after another, they pollinate the plants.

Kingbirds

Gray Kingbirds are very visible local flycatchers, larger than the Caribbean Elaenias. They sit on trees or wires, and fly out from their high perches to grab bugs they see in the air, then fly back to the same spot. They have a loud sound like someone blowing a whistle, and as a result have been called 'chi-chery' birds.

Parakeets

Brown-throated Parakeets (also called St. Thomas Conures) are squawky birds that travel in small flocks looking for seeds, fruits and flowers in trees. Some of them ended up in Cruz Bay after the 2017 hurricanes and they are now spreading out across the island. They originally were brought to the Virgin Islands from South America in the 1800s.

Pigeons

Rock Pigeons are originally from Europe and Asia. People brought them to North America to raise them for racing, carrying messages, or for food. A small flock is living wild in Cruz Bay. They generally eat seeds, grains, fruits and insects, but will eat people food also, like bread.

Scaly-naped Pigeons are local birds that gather in trees and on the overhead wires. They have bluish-purple feathers, dark red eyes surrounded by yellow, and reddish necks. They mostly eat fruits and berries. They make nests in tall trees, including sometimes in Francis Bay. You might hear them calling to each other there: *"who, hoo, hoo, hoooo".*

Sapsuckers

Yellow-bellied Sapsuckers are uncommon but exciting winter visitors. They are woodpeckers with bright red heads, and they feed themselves by drilling a series of holes in circles around the bark of large trees. As sap collects in the holes, they slurp it up with their long tongues, along with any insects that were attracted to the sap. So if you see a tree with a row of holes in a straight line, you'll know a sapsucker probably made it. (Look along the loop trail across the road from Cinnamon Bay.)

Sparrows

House Sparrows can be seen living in a few areas of Cruz Bay and Coral Bay. They hang out in groups in trees, chirping loudly. They are primarily seed and grain eaters, and were originally brought over from Europe. When they were introduced into North America, they spread out widely, mostly in cities. They might have come over to the Virgin Islands riding along on ships.

Swallows

Barn Swallows sometimes pass through St. John as they move between their breeding areas in North America and their winter habitat in South America. They have bright blue backs, with orange on their throats and faces. They are known for their forked tails, which help them maneuver in the air to catch insects.

Thrashers

Pearly-eyed Thrashers or 'thrushees' are aggressive local birds that can be seen all over St. John, and other islands in the eastern Caribbean. They will eat a lot of different things - fruits and berries, large insects, frogs and crabs, and even other birds' babies. And don't leave your lunch unattended at the beach or they will get into it. They make nests in trees, on buildings, and in the high rafters of the park pavilions.

Shore and Ocean Birds

Boobies

Brown Booby birds are residents that nest on offshore islands. They are easy to see at the beaches, as they fly low across the water to grab fish, or dive down from the air. Or you might see them sitting out on the buoys. The adults have white faces and bellies, brown wings, and yellow feet. Sometimes their bellies look blue as they fly close to the water and pick up the reflection. The young ones are brown all over, with blue bills and pinkish feet.

Frigatebirds

Male

Female

Magnificent Frigatebirds hover high overhead, and are easily recognizable by their sharply arched wings (which can be over 6 feet across) and their forked tails. They are known as 'weather birds' because they come inland during storms. They can stay in the air for weeks, catching warm updrafts of air and then riding the wind. The updrafts also churn up the water and bring fish to the surface, where the frigatebirds can grab them. But they can't dive in, or they will drown. Their feathers are not weatherproof and get very heavy when wet. An alternative way the frigate birds get fish is by stealing them from other birds, which has earned them a reputation as pirates. The females are black with white heads and chests. The males are all black except for red pouches on their necks that they can inflate to attract the females. They make nests on a few offshore cays in the British Virgin Islands.

Gulls

Laughing Gulls arrive in St. John around the first of April and stay for the summer. In late fall, they disperse around the area. Besides diving for fish, the laughing gulls will also walk along beaches and shorelines looking for shellfish, crabs and large insects. They come to make their nests on the cays, which provide some protection from predators.

Kingfishers

Male

Female

Belted Kingfishers are winter residents that perch on bare branches and stare down into the shallow water to look for small fish. They dive down quickly to grab the fish, then return to their perches, maybe letting out a loud "*chee, chee, chee*" sound as they fly. They have wide blue bands across their bellies, and the females have rusty patches there as well.

Ospreys mostly come in the winter, and leave early in the spring to get their nests ready up north, though a few have occasionally stayed over in this area through the winter. They are also known as 'fish hawks', because that's all they eat. They have very sharp eyesight so they can fly high and spot fish in the water below. Then they dive straight down, underwater, to snatch up a fish with their strong talons.

Oystercatchers

American Oystercatchers live on rocky, isolated beaches on the south shore of St. John, and on some nearby offshore cays. They are dramatic birds, with large bright, reddish-orange bills, dark heads and backs, white bellies and pink legs. There are usually a male and female together. They make shallow nests scooped out in the sand, and mostly eat shellfish that they find along the shoreline. They use their strong bills to pry open the shells, and locally are known as 'whelk crackers'.

Pelicans

Brown Pelicans are visible most of the year at the beaches. They move around the area depending on where the fish are. It can be very relaxing to sit and watch them dive repeatedly into the water. They might go after a particular fish, or scoop up a pouch full of water and swallow the small fish left inside after the water drains out. Their nests on trees by Mary Creek were blown down during the 2017 hurricanes, but they still nest on offshore cays.

Plovers

Black-bellied Plovers pass through in the winter or early spring, mostly on the rocky south shoreline and tidal flats. They spend their time searching for worms, insects and small shellfish. The males grow black feathers on their faces and bellies to attract females when they fly north to mate, but on St. John they are pale.

Semipalmated Plovers are the most common ones on St. John, and will occasionally turn up on the beach at Francis Bay. They generally run quickly along the edge of the water looking for small shellfish and marine worms. Their name 'semipalmated' means that their feet are partially webbed. They have yellow legs, and around their necks there are black lines like necklaces. They fly to Alaska to nest.

Wilson's Plovers are resident birds, not migrants, and make shallow nests on the ground. They have thick bills that they use to break up fiddler crabs and snails and other small shellfish.

Turnstones

Ruddy Turnstones get their name from their reddish color and their way of foraging for small shellfish, worms and insects as they work their way along the shoreline. This photo was taken in the spring, when the birds were already growing their dramatic mating feathers, and preparing to fly up to the Arctic to mate and raise their chicks.

Sandpipers travel south from their breeding grounds in Canada and Alaska for the winter to feed along the coasts and marshes in the Caribbean and South America. Some of them show up on St. John.

Least Sandpipers are the smallest of the sandpipers you might see on St. John shorelines, but they are not very common. They mostly pass through going south in the fall, then back north in the spring.

Semipalmated Sandpipers are often visible on St. John, particularly in the fall. They are called 'semipalmated' because their feet are partially webbed.

Spotted Sandpipers are quite common in the ponds on St. John during the winter, including the one at Francis Bay. They are usually alone, not in a group. The name is confusing, because during the winter they don't have their spots. But in the spring before they head north, the spots start appearing. One easy sign that it's a Spotted Sandpiper is the way it bobs its body up and down.

Stilt Sandpipers regularly stop by St. John especially in the spring. They generally move around in close, small groups. They are relatively large, for sandpipers, with long legs. Their distinctive characteristic is the way they move their heads and bills up and down while feeding - in what is called a 'sewing machine motion'.

Terns are fish-eating sea birds that are summer migrants to the St. John area, where they nest on offshore cays.

Least Terns have yellow bills with black tips, and yellow feet.

Roseate Terns have black bills and red feet.

Royal Terns have orange-red bills and black feet. They are the largest and most common terns on St. John, and can often be seen sitting on buoys at the north shore beaches.

Sandwich Terns have black bills with yellow tips, and black feet.

Pond and Wetlands Birds

Coots

American Coots dive in ponds, feeding on aquatic plants, fish, and insects. They have interesting lobed feet that help them move powerfully through the water. Some live all year on St. John, and others come in the winter from North America. They have white bills, and connected 'shields' just above their bills. Sometimes these shields are dark red, and sometimes white, like a continuation of their bills. The birds with white shields were once thought to be a separate species of 'Caribbean Coots' but are now considered variations in color within the same species.

Ducks

Blue-winged Teals are common ducks in St. John ponds during the winter, especially by Francis Bay. Then in the spring most of them go north to nest. The males have white crescent shapes on their faces, but the females do not. Both have blue patches on the wings, which you only see when they fly or raise their wings. These are 'dabbling' ducks - they usually put their heads down and grab vegetation under the water, sticking their butts up in the air.

Male

Female

Ruddy Ducks on St. John are mostly winter migrants from North America, though there are a few that spend the whole year in this area. The males attract the females with their reddish-brown feathers, plus their dramatic blue and white bills. The females are plainer, mottled brown with darker heads. Both have tails that stick up in back. They will dive underwater to forage for insects and tiny shellfish, and can stay out of sight for a long time before resurfacing.

Male

Female

White-cheeked Pintail Ducks are common resident birds in St. John, and are also found in other parts of the Caribbean and South America. They are dabbling ducks, often just reaching down to get vegetation on the bottom of the pond, although they can go all the way under if the water level in the pond is too high for dabbling. They nest on the ground not far from the pond, and when the chicks have hatched you can see a group of them trailing across the pond guarded by their parents.

Egrets

Cattle Egrets live all year on St. John, mostly around Coral Bay. (Egrets also live in many other places around the world.) Here the cattle egrets roost and breed in the mangroves near the water, but their diet is mainly land-based, including insects, lizards and frogs. They get their name because they follow farm animals around and grab insects and other creatures that get pushed up while the animals are grazing. They have shorter necks than the other egrets.

Great Egrets on St. John are a mixture of permanent residents and winter migrants from North America. You can often see them fishing in and around the ponds, or along the shoreline, but they are probably most visible when they are staring into the bushes along the roadsides, hunting for lizards. They are the largest of the egrets here (up to about 3 feet tall), with long yellowish bills and long black legs. They were once hunted for their frilly tail feathers, which were then used to decorate ladies hats. The public campaign to protect them, and Snowy Egrets, led to the formation of the National Audubon Society in 1905.

Snowy Egrets were also hunted for their attractive feathers. They can be found in a wide range of places in North and South America and the Caribbean. Although some may be residents of nearby islands, there are not very many visible on St. John, and those seem to leave for the summer. They are smaller than the Great Egrets, with black bills and yellow on their feet and legs, and have a more energetic hunting style. In the ponds, they move around quickly in the water, kicking up fish, crabs and insects, then striking with their bills using short, stabbing motions.

Gallinules

Common Gallinules (previously called Common Moorhens) are similar to Coots, but with bright red bills and shields on their faces as adults. They also have yellow on their legs, feet and the tips on their bills. They live here all year, mostly eating plants growing in and under the water. They make visible nests along the edges of ponds.

Grebes

Least Grebes live all year in some of St. John's ponds, making their nests on piles of weeds in the middle of the water. They are small, as their name indicates, but are not shy. They have bright, bold yellow eyes, and make loud peeping sounds to let you know they are there. Then they will quickly dive underwater and disappear, looking for aquatic insects or small fish to eat.

Adult left and juvenile right

Parent on nest

Pied-billed Grebes are larger and more secretive, living farther back in hidden ponds. They have dark marks across their bills.

Adult

Juvenile

Herons

Great Blue Herons can get up to 4 feet tall, almost twice as big as Little Blue Herons. Most of them come here from North America just for the winter. They fish in ponds and wetlands, and also along the shoreline. They stand quietly and then stab a passing fish, grab a crab, or catch a rodent or bird.

Green Herons have a lot more colors on them than their name would suggest. They live all year on St. John, and nest in the wetlands. They are territorial about their ponds and squawk aggressively at intruders even though they are quite small. They are usually seen fishing, in a pond or occasionally along the shoreline, but will also eat crabs and large insects. They will stand still, peering into the water, then quickly extend their long, rusty colored necks to grab their prey.

Little Blue Herons are tricky because they are not always blue. For their first year they are white, and then as they mature the blue feathers gradually come in. Although some of them do come to St. John only for the winter, there are others that stay year-round and nest here. They are often in the ponds, and also show up along the shorelines, slowly looking around for fish. They will also eat crabs and insects and lizards.

Adult

Juvenile

Night Herons

Black-crowned Night Herons live on St. John all year but stay in hidden ponds. They live in many other areas of the world as well, though, and are not always so shy. The adults have white bellies, dark backs and caps, and black bills. They also tend to hunch over, and some people think they look a bit like penguins. The juveniles have brown and white stripes on their bellies, patterns with white points on their backs, and bills that are bluish on top with yellow on the lower part. One of the reasons that they come out at night is that they like to eat the land crabs, which are active after dark.

Yellow-crowned Night Herons are also resident birds that are out in the evening hunting for land crabs. They can sometimes be seen on the roads along wetlands areas, probably because they find crabs out there too. During the day they rest in the mangroves. They are a bit larger and more upright than the black-crowned ones, with heavier, blunter black bills. The white patches on their faces are more noticeable than the pale yellow coloring on their heads. It can also be quite difficult to tell the difference between the juveniles of the black- and yellow-crowned ones when they are creeping around and camouflaged in the mangroves.

Adult

Juvenile

Tricolored Herons have blue heads, necks and backs, but their bellies and 'underparts' are white. Some people have read about them in books and thought it said 'white underpants'. That can actually be a good way to tell them apart from Little Blue Herons, which are about the same size. The Great Blue Herons are quite a bit taller. There are not many Tricolored Herons on St. John, and it seems that they are mostly around during the winter.

Rails

Clapper Rails get their name because they make an incredibly loud clacking sound - *kek, kek, kek, kek, kek* - so loud you think there is a whole flock, though it might only be a couple. They are permanent residents, but secretive so you won't see them very often. They creep along the edges of ponds looking for crabs, worms, little fish and insects, and make their nests in hidden areas around the ponds.

Soras are winter residents in several St. John ponds. Though they can make loud, screechy calls, they are usually just quietly walking around the edges of the ponds looking for insects, seeds or little snails. They are hard to spot because they are usually camouflaged by vegetation. Their feet and bodies look similar to the Clapper Rails, but they are smaller and have shorter necks and bills than the clappers.

Stilts

Black-necked Stilts are very common year-round residents in St. John's ponds, usually gathering in noisy groups. They have eye-catching black and white feathers and long, thin pink legs. They wade into the water to pick up small fish, shellfish, worms and insects, and make nests on piles of sticks and weeds.

Whimbrels are migratory shorebirds that breed in North America and winter in the Caribbean and South America. There was a famous one named Hope, who for many years wore a satellite transmitter documenting her annual travel to Great Pond on St. Croix. Although whimbrels are not common on St. John, they are occasionally spotted by ponds on the south shore.

Yellowlegs

Yellowlegs are common winter migrants to St. John ponds. They wade into the shallow areas with their long legs, or stand around the edges, picking up fish, little crabs and insects. They come in two varieties, which look very similar and are frustratingly difficult to tell apart.

The **Greater Yellowlegs** are a little over a foot tall, several inches more than the **Lesser Yellowlegs,** and have slightly longer, thicker bills and legs. Their heads and bodies are also a bit bigger, but it is really not that obvious when you only see one of them.

Greater

Lesser

Warblers

The term 'warbler' includes a variety of small songbirds, most of which feed on insects. They generally save their songs for breeding season, and just make 'chip' sounds when they are here. If you hear one chipping, it can take a lot of patience to keep looking until you finally see it. They might be around Francis Bay, across from Cinnamon Bay or in other wooded areas.

Some warblers come from North America and stay for the winter. Others just pass by during their annual migration between wintering areas in South or Central America and breeding grounds in Canada, Alaska or the northern United States.

There is one type - the Yellow Warbler - that is a permanent, non-migratory resident. Another type now living here full time is the Adelaide's Warbler, which used to be only in Puerto Rico, but recently expanded its range to St. Thomas, and then to a remote part of St. John's south shore.

American Redstarts are common winter residents. The males are black with orange-red patches, while the females are more yellow and black, with a bit of orange. They flit around in the trees flaring their tails and wings as they chase bugs.

Male

Female

Blackpoll Warblers, both male and female, are a mix of light green and black in the winter when they stop by here. In the summer, when they're not here, the males have stylish black caps, which is how they got their name.

Winter

Male in summer

Black-and-white Warblers stay around St. John for much of the winter. Their strategy is to crawl along a tree's trunk or branches, looking for bugs, so they are easier to spot than the warblers that are in the leaves on treetops.

Northern Parulas are colorful warblers that can be seen through much of the fall and winter, often quietly chasing moths and other insects in the tops of trees. Other times they make 'chip' sounds, so you notice them. Their backs are blue, with a wide green spot, and they have white wing bars on their sides. Their bellies are white, and their throats are yellow.

Northern Waterthrushes winter regularly in the wetlands of St. John. Unlike many of the other warblers, these birds creep around low down in the wetlands, where they hunt for insects and small crustaceans. They make a loud 'chip, chip' sound, which lets you know that one is close.

Ovenbirds are not very common on St. John, but have shown up occasionally, including on the trail across from Cinnamon Bay. They are usually on the ground, poking through the leaf litter for insects, bobbing their tails a bit. They have chest stripes like Black-and-white Warblers and Northern Waterthrushes, but different foraging patterns. Look for a bit of orange on their heads.

Prairie Warblers appear occasionally in the winter. They have yellow chests and bellies, and greenish backs, a bit like the Yellow Warblers. However, they have distinctive dark marks on their faces, like black eyeliner.

Prothonotary Warblers are beautiful but rare on St. John. They have bright yellow heads and bellies, blue wings, and a greenish area on their backs. They have been spotted in the spring next to ponds along the north shore.

Yellow-rumped Warblers are sometimes on St. John in the winter, and easy to identify if you see them from the back. In general, though, they are not very noticeable. They like to eat spiders, and for that diet they seem to prefer isolated swamps or forest areas.

Yellow Warblers are permanent residents of St. John, usually found near the mangroves and wetlands. They are bright yellow with greenish wings, and the males have rust-colored stripes on their chests. They are sometimes called 'canaries' because they are so yellow. When they sing, some people think it sounds like they are saying: *"Sweet, sweet, I'm so sweet"*.

Local Celebrities

Flamingos

American Flamingos used to be common in the Virgin Islands, but they died out not too long ago, probably due to a mix of hunting, hurricanes, predation and loss of habitat. In the British Virgin Islands, a small flock of flamingos was reintroduced in Anegada in 1992, and then in 2006 another group was brought to Necker Island by Richard Branson. Those flocks have been breeding and growing, and some birds have spread to other parts of the BVIs, with occasional visitors to St. Thomas and St. John.

It takes the young ones several years to get their full color (with a proper diet) and in the meantime they have black and white feathers, and pale pink legs and throats. Their pink coloration as adults comes from carotenoids in microscopic algae, which flamingos are able to metabolize to create pigments in their feathers. They can eat the algae directly, or get the carotenoids from eating brine shrimp or other tiny creatures that also feed on the algae growing in shallow, salty ponds.

Ibises

Scarlet Ibises appeared on St. John in 2019, and one began living in the Fish Bay area. They are native to Trinidad and Tobago, and northern Venezuela, not the Virgin Islands. They likely came over as wanderers from a flock that was introduced to Necker Island in 2010, as part of Richard Branson's collection of birds and endangered animals. Ibises are amazingly bright red when they are adults, with black wing tips. They are paler and more pinkish, when they are young. Like flamingos, they get their color from carotenoids in their diet. They push their long bills into the mud or shallow pond water, and poke around to find small shellfish and crabs.

Checklist and Index **(Common names plus scientific names in Latin)**